ON WARRIORSHIP
A Guide for Your Warrior Path

By Todd K. Hulsey
USMC & USAF

To the warriors, past, present, and future.

Printed in the United States of America

Hulsey, Todd K., 1964-
On Warriorship / Todd K. Hulsey

Includes bibliographical references and index.
ISBN 979-8-9885019-0-9 *Hardcover*
ISBN 979-8-9885019-1-6 *Paperback*
ISBN 979-8-9885019-2-3 *E-book*
ISBN 979-8-9885019-3-0 *Audiobook*
1. Warriors.   2. Soldiers.   3. Military arts and sciences.   4. Sociology, Military.   5. History, Military.   6. Philosophy.   7. Leadership.   8. Martial arts.   9. Self-help.

Cover art by William Black of Black Draws Stuff.

Cover design by Velmacover.

Audiobook narrated by Jon-David Wells.

10 9 8 7 6 5 4 3 2 1

CONTENTS

DEDICATION

This book is dedicated to my lovely wife Tammy, my daughters Emma and Leah, my parents and brother, and all those friends, teammates, colleagues, leaders, and mentors without whom I would not be me.

I also dedicate this book to Lt. Colonel Kent Kershenstein, USAF (Ret.), commander, mentor, and friend, who taught me more about officership than anyone, and who is one of America's finest officers. Thank you.

A special thanks to you all. Semper Fidelis!

PREFACE

Arguably, conflict – warfare – is the quintessential human activity. Thucydides, the ancient Athenian general and historian, called war "the human thing." An honest reading and assessment of human history reveals that this is so. Some might disagree. However, I challenge doubters to establish when human interaction has been devoid of conflict, from the interpersonal to the nation-state. Great civilizations have been born of warfare, just as great civilizations have died by warfare. War is omnipresent on Earth.

Great technological advances have resulted from warfare. Take aviation, for example. The first verifiable flight of a powered, controlled, heavier-than-air vehicle was on December 17, 1903, at Kill Devil Hills, North Carolina.[1] On July 20, 1969, less than 66 years later, the Apollo 11 moon mission landed Neil Armstrong and Buzz Aldrin on the moon (with command pilot Michael Collins orbiting the moon in the command module)

---

[1] This is usually reported as Kitty Hawk, North Carolina, but the actual flight was at Kill Devil Hills.

This rapid advancement in aviation from wooden, fabric-covered aircraft to spacecraft was the result of two world wars and the Cold War between the United States and the Soviet Union.

Indeed, warfare is one of the greatest drivers of civilization, perhaps *the* greatest. Great and terrible things stem from warfare, but without it, the world we live in would be a very different place. Whether it would be a better or a worse place is unknown, but it would be very, very different.

Warfare as a human activity means that humans are the active component. Warfare necessitates the warrior. There are, have always been, and always will be, warriors and warriorship.

With the advent of drone warfare and artificial intelligence (A.I.) technologies for use in war, the nature of warriorship may change in some ways. Such technologies may remove humans from the battlefield to an extent not previously seen in warfare. To what extent and how this may change warriors and warriorship, if at all, cannot be clearly foreseen.

But, so far as human history teaches, the methods of war may change, but the attributes of the warrior tend to stay constant. With remote weaponry such as drones, someone must still "pull the trigger," so to speak.

Truly autonomous weapons driven by A.I. technology must still be programmed by someone, and the decision to deploy these weapons must be made by someone. The human element will remain somewhere in the proverbial "decision loop."

Warriors and their warrior technologies are as different as the time and place in which they were called to be warriors. But the warrior is also the same across time and place. While this may seem like a paradox, it really is not. While the kinds of war and the ways they are fought change over time, the type of person who takes up warriorship does not. The tribal warrior emerging from the last Ice Age was made of the same stuff as are warriors today. Wars change; warriors don't.

The warrior is necessary to fight wars. The warrior is also necessary to "fight the good fight" in solving mankind's intractable problems, big and small.

Yellow fever was once considered an intractable problem, ravaging the subtropics. This often deadly disease primarily affected Sub-Saharan Africa and South America, stifling human activity and economic development. Those not killed by the disease usually suffered lifelong debilitation due to liver damage.

In 1900, U.S. Army physician Dr. Walter Reed led a team of Army researchers to Cuba to study infectious diseases, particularly yellow fever. His team confirmed that it was transmitted by a certain species of mosquito, as hypothesized by Cuban physician Carlos Finley. Reed's work led to mosquito eradication efforts in the tropics and laid the groundwork for development of the yellow fever vaccine.

While not a warrior at war, Major (Dr.) Reed was a warrior against infectious disease and is credited with laying the groundwork for the global eradication of yellow fever.

Whether at war in combat, or at war with some of mankind's most difficult problems, warriors serve causes greater than themselves, and do so with honor.

ACKNOWLEDGMENTS

The genesis of this book was an invitation from Mr. Randel Davis, host of the *Kung Fu Conversations* podcast, to participate in a multi-podcaster, multi-podcast conversation on warriorship. Randel's vision was to gather the various opinions of different martial arts podcasters regarding this topic to better understand the concept of warriorship. My podcast, *The Dirt Wasp Podcast*, was honored to participate. The objective was to develop multiple points of view from those with a shared interest in a warrior-related subject – martial arts. Comparing these viewpoints would help ascertain a better understanding of what "warriorship" really is, and what it is not. Without the vision of Randel Davis, this book would not have been written.

I also want to thank my wife Tammy Hulsey, novelist Patricia McFerren, GySgt. Michael Lepek, USMC (Ret.), business executive and former Marine Gary Sirkel, my former FBI boss, mentor, and friend, David W. James, and educator Michelle Connelly for their critical feedback on the manuscript. And to my book editor, Christopher Crompton, thank you.

FOREWORD

The world emerged from a global crisis forever changed. Some of this change was for the better, some for the worse, and many people were left looking for answers in their lives. On the positive side, some people found gyms, yoga, martial arts and many other physical outlets. Others turned to religion, ideals, philosophy and culture. It could be argued that the oldest and most foundational of all of these is the Warrior Culture.

So, what defines a warrior? Is it tangible, touchable, reproducible? Does it involve certain qualities, characteristics, principles or morals? In seeking to answer these questions, we should consider who the most appropriate people are to ask about warriorship and what it takes to pick up the mantle of the warrior.

Hollywood might have you believe that a warrior is Rambo, shooting down a Russian helicopter with a bow and arrow, or Arnold Schwarzenegger swinging his sword and axe, fighting the horde alone. As you dig deeper into this book, you will find that the warrior rarely works alone. Even the concept of the "lone wolf" is a false image, as the outcast wolf dies quickly when left to survive on its own. If not to Hollywood, should we look to artists, poets, or politicians to define warriorship? Or should we instead ask those we might consider the warriors of the present day?

The book you're holding right now has been penned by a man who has taken up the title of warrior and whose life has been shaped by the art of warriorship. Finishing thirty-nine years of United States military service, as well as spending more than twenty years as a federal agent, Todd Hulsey is a man who has served his country and fellow man by carrying his duties of service with honor and integrity.

Working with others in foreign countries and stateside to keep his fellow Americans safe is warriorship in action. Using The *Dirt Wasp Podcast* as a platform to reach out to the like-minded modern-day warrior, and being a student of the history of warrior cultures, the Colonel has expanded outside of his own experience to foster a broader understanding of warriorship.

Whether you have walked the Warrior's Road for a lifetime, or are just about to take your first steps, when opening this book, you take part in a dialog and mindset that started at the beginning of mankind and will be there in the final days.

~ Randel Davis
Kung Fu coach, podcaster, bus
driver, and mountain preacher.

I am so very honored to write this foreword for Col. Todd Hulsey and his book, "*On Warriorship.*" I had several key male role models growing up. The first was my grandfather, who raised me from nine months of age until I went into the military. He was a World War I veteran. Two of my uncles served in World War II, both going on to serve for over 20 years before retiring. My younger uncle served four years in the Marines during the Korean War. They instilled the warrior belief and creed in me from an early age.

I served for 22 years and 7 months of active duty in the U.S. Air Force and retired as a SMSgt (E-8). After my Air Force career, I worked for General Dynamics/Lockheed Martin and retired after 23 years of service. I then joined the Texas Guard and served for 5 years and 7 months, retiring as a CMSgt (E-9).

During my time in uniform, I served under numerous commanders in numerous locations – Vietnam, Korea, Thailand and deployments throughout Europe. Col. Hulsey is one of the best commanders that I had the honor to serve under! He always took care of his troops, the mission, and our honor! His word was his bond. We all trusted him without hesitation, not just because he was our commander but because he had been there and done what we were tasked to do! He was and is a warrior of the highest level.

There are two types of warrior:

1.   The first type is the warrior in military uniform. This is a warrior bound by the oath that we all pledge to live by, and by our warrior creed. As a team member or team leader, it is our mindset and heartfelt responsibility to take care of all of our brothers and sisters who stand and serve beside us. Some call them wingmen or battle buddies. They are our team!

2.   The second type of warrior is the civilian (community) warrior. This is a person who may or may not have served in the military. They watch out for their neighbors and citizens in the community who need help of any kind and take action to help them! Both of these types have the same type of warrior spirit! They believe in selfless service, "service before self!"

Col. Hulsey has detailed the warrior creed in simple terms. This is an excellent read! Well worth your time! The warrior is bound to take care of their team, take care of their mission, and come home with their honor!

"GET ER DONE!!! BROTHERS FOREVER!!!"

~ Robert "Guns" Moore
SMSgt, USAF (Ret.) & CMSgt,
Texas Guard (Ret.)
Three combat tours, Vietnam War

INTRODUCTION

Warriorship: the condition or state of being a warrior.

This purpose of this book is to add additional perspective to the definition of warriorship. If you picked up this book, you most likely had a working definition in your own mind as to what a warrior is and are interested in learning more about warriorship.

This is not an academic study on warriorship. This is not an exhaustive recitation of the available subject matter presented through my own lens. This is a discussion, a conversation, food for thought. It is written concisely in plain, simple language to make it accessible to a wide audience.

Each person will have a different definition of what a warrior is and is not. Through our comparison and contrast of these definitions, we can better come to a consensus on what a warrior is in today's society.

Some people will ask the question, "Why be a warrior?" Or, put another way, "What is the purpose of a warrior?" This is an important question. There have been warriors throughout history, all over the world, and there are warriors today. But why?

Your answer to this question may be different from mine. Warriors are necessary to fight wars, simple and true enough. But is that all a warrior is good for, is that all a warrior does? I believe that the answer is broader than that.

Being a warrior is, above all, about service to others.

> "The warrior, for us, is one who sacrifices himself for the good of others. His task is to take care of the elderly, the defenseless, those who cannot provide for themselves, and above all, the children – the future of humanity."
>
> — Sitting Bull (Hunkpapa Lakota leader)

There is likely no such thing as a definitive understanding of warriorship. War and warriors are as diverse as human needs and imagination. "War on Poverty," "War on Drugs," "War on

Terror," "War on Cancer," World Wars I and II... There are so many variations on "war" that the type of "warrior" is almost as varied, depending upon time, place, and necessity.

Is a cancer researcher who is fighting against cancer a "cancer warrior?" Is a soldier necessarily a warrior, or can they be "just a soldier?" You can see how this can get complex in a hurry!

Discussing warriorship and trying to get down to a single, definitive definition is a lot like peeling an onion. Each of those peels contain a bit of information about warriors and warriorship. There is layer upon layer to peel away until you get to the core.

And the core is only the center of the onion; it is not the sum of it. The center is devoid of all those layers that have been peeled away. Even it lacks the total meaning of this onion called warriorship. There no right or wrong answer, only differing perspectives.

The basis for this book is my thirty-nine years of military service and my twenty-plus years of federal law enforcement and intelligence service. My training informs my opinion on warriorship, and my experience sharpens my understanding. The conclusions I draw are certainly opinion, but well-informed opinion.

This is an American-centric book because I am an American. But warriors and warriorship are universal. This book is also the product of its time and place, being written in the early first-half of the 21st Century in the United States of America. But warriors and warriorship are timeless.

I am hopeful that this book adds to the discourse on this topic and helps you develop your personal perspective on warriorship and helps you with walking your Warrior Path.

CHAPTER ONE

WHAT IS, AND WHAT IS NOT, A WARRIOR?

How often do we hear a sportscaster or commentator exclaim that a player, or a boxer, or an MMA fighter is a warrior? "He's a real warrior out there!" Or they describe an upcoming game or match as "war:" "It's gonna be a war tonight!"

They are trying to say that the competitor has the spirit to never give up, that the game is going to be hard fought. But are war and warrior accurate descriptions or just easy metaphors?

In my mind, this is metaphor, plain and simple. An MMA fight or a rugby match is not warfare, and those who compete are not warriors. The competitors may have a "warrior's heart," or the "warrior spirit," but they are not warriors. At least not in my opinion, and that opinion will fill the rest of these pages.

Warriorship is the "condition or state of being a warrior," which is our working definition. However, we must then define what it is to be in the state of being a warrior. It is not easy to define what "warrior" means. Let's explore it. There is no absolute definition and, of course, there are different ways of thinking about the same thing.

We are all products of our time and place. How we define a warrior and warriorship is dependent upon the time, place, culture, and circumstances of the society that produces the warrior. For example, a Roman legionary and a Comanche brave are both warriors but are also so different as to almost defeat comparison. So, let's turn to what a warrior is not.

A fighter is not a warrior. A fighter just fights, whether a street fighter, a bar fighter, a sport fighter, or a prize-fighter. This does not mean that a fighter lacks a "warrior spirit" or a "warrior mindset," but those things alone do not make a warrior.

A predatory criminal is not a warrior. He may fancy himself a warrior, and he fights (often from ambush), but he is not a warrior. A brigand fights but is certainly not a warrior.[2]

---

[2] Brigand: a person who lives by pillage, plunder, and robbery. Merriam-Webster Dictionary, 2022.

A martial artist is not a warrior. A warrior may study martial arts, may practice martial arts, but a martial artist is not inherently a warrior.[3] Neither is a military member necessarily a warrior. For example, although contributing greatly to the mission, a military computer programmer who works in a safe environment, even during wartime, is not necessarily a warrior, even though his or her work is essential to the war effort.

My definition of a warrior is this:

A warrior is a person who is trained for and prepared for engaging in that enduring, quintessential human activity that we call war; one who is intellectually, emotionally, and spiritually ready to engage in warfare in all its forms but most particularly combat on the battlefield, whether it be on land, sea or in the air or space; and who is capable of taking human life and sacrificing their own in service to something greater than themselves.

---

[3] Much like the definition of "warrior," there are several possible definitions for "martial arts" and "martial artist." In the context here, martial arts/martial artist refers to anything and anyone who trains in a dojo, dojang, academy, gym, or school that bills what it does, and what they do, as "martial arts."

Take note of the last part: sacrificing in service to something greater than themselves. This is the most important aspect of being a warrior.

My definition is born out of who I am and what I have done. I am a lot of things in this life, but the rock-bottom base of my being is that I am a soldier. My foundational experience is my service in the United States Marine Corps.

I am admittedly biased toward the military warrior. But does my definition exclude someone who is not trained for and prepared for war? Not necessarily, and we will get to that in Chapter Eight.

CHAPTER TWO

THE ATTRIBUTES OF THE WARRIOR

The attributes of the warrior are timeless. These attributes tie the ancient warrior to the modern warrior. They are constant from the days of the pre-historic tribal warrior, on to the Sumerian army and still onward to the present day, and all that has existed in between. These attributes are having a warrior spirit, moral courage, ethical behavior, physical courage, physical preparation, selflessness, accountability, humanity, fidelity, and foremost, honor.

THE WARRIOR SPIRIT

The warrior spirit, or warrior mindset, (essentially the same thing, in my view) involves having the gumption to never quit, to carry on in the face of daunting adversity – to overcome the odds, to keep fighting, to persevere. The warrior is imbued with the warrior spirit.[4]

---

[4] There are hundreds of books, articles, podcasts, and speeches that talk about the "Warrior Mindset" and the "Warrior Spirit." I recommend the reader seek out these various sources for more information on this particular topic.

"The basic difference between an ordinary man and a warrior is that a warrior takes everything as a challenge, while an ordinary man takes everything as a blessing or a curse."

> – Carlos Castañeda (Peruvian-American writer and thinker.)

MORAL COURAGE

Warriors seek to understand their society's moral foundation and principles. With this understanding of right and wrong, they strive to do what is right, even when it is difficult to do so. The warrior stands up for what is right and denounces wrongdoing, even in the face of criticism and derision. The warrior speaks and lives the truth.

"To see the right and not do it is cowardice."

> – Confucius (Ancient Chinese philosopher)

PHYSICAL COURAGE

The warrior is willing to face personal danger and engage in battle. The warrior does not run from danger, but, when necessary, runs toward it.

Courage is defined as the mental and emotional strength to venture, persevere, and withstand danger, fear, or difficulty.[5] Most people, including warriors, focus on physical courage. But moral courage is more difficult to muster. The warrior must have both.

To be discouraged is to have that mental and emotional strength taken away. To be encouraged is to find that mental and emotional strength. Warriors seek to encourage themselves, and to encourage their fellow warriors. They never seek to discourage themselves or their fellows.

> "Courage is resistance to fear, mastery of fear, not absence of fear."
>
> — Mark Twain (American writer and humorist"

---

[5] Merriam-Webster Dictionary, 2022.

ETHICAL BEHAVIOR

Human society worldwide likely shares a type of "universal morality," a universal ethical standard. These shared ethical standards have been described as the "Seven Types of Morality" – obligation to family, loyalty to group, reciprocity (helpfulness), bravery, respect, fairness, and respect for the property of others.[6] Indeed, each society that produces the warrior has its own mores, sense of morality, and understanding of what is right and wrong. Warriors strive to act in accordance with the moral and ethical framework of their society and aspire to exceed their society's expectations of them.

"I would prefer to fail with honor than win by cheating."

– Sophocles (Ancient Greek writer and playwright)

---

[6] *Current Anthropology*, "Is It Good to Cooperate? Testing the Theory of Morality-as-Cooperation in 60 Societies," Oliver Scott Curry, Daniel Austin Mullins, and Harvey Whitehouse, University of Chicago Press, Volume 60, Number 1, February 2019.

PHYSICAL PREPARATION

The warrior is physically fit, skilled, and tough. The warrior is physically prepared for battle. The warrior works to achieve and maintain this preparedness.

> "Physical fitness is not only one of the most important keys to a healthy body, it is the basis of dynamic and creative intellectual activity."
>
> – President John F. Kennedy (35th President of the United States)

SELFLESSNESS

The warrior seeks to serve others before themselves. The warrior serves something greater, whether this is a cause, a purpose, a group, or a nation. This service is lawful; it is not seditious, treasonous, or criminal. The warrior is willing to sacrifice.

> "True greatness, true leadership, is achieved not by reducing men to one's service but in giving oneself in selfless service to them."

– J. Oswald Sanders (New Zealand lawyer and missionary)

ACCOUNTABILITY

A warrior is willing to be held accountable by peers and the society the warrior serves. Warriors strive to meet or exceed the standards of their peers and of their society. Should they fall short, they take complete responsibility for their action or inaction and submit to the judgment of their peers and society. The warrior is humble, self-aware, and dedicated to self-improvement.

> "Accountability is the glue that bonds commitment to results."
>
> – Bob Proctor (Canadian writer and thinker)

HUMANITY

The warrior does not lose his or her humanity, even in the darkest throes of battle. The warrior respects the enemy and treats the vanquished foe with humanity and respect, even if the foe itself lacks humanity. This is very hard to do, but essential nonetheless.

"In recognizing the humanity of our fellow beings, we pay ourselves the highest tribute."

— Justice Thurgood Marshall (Associate Justice, U.S. Supreme Court)

FIDELITY

Warriors are faithful to their purpose and the cause they serve. They are loyal to their fellows and develop a bond of trust with them. The warrior is part of a team, so must be a team player and a good teammate. Warriors put the team above themselves, living, eating, sleeping and fighting as a team. Life is a team sport!

"Nothing is more noble, nothing is more venerable than fidelity."

— Cicero (Roman statesman)

HONOR

The warrior is honorable. Warriors honor their society, their fellows, their purpose, their work, and even their foes. They seek to never stain his honor, and to never stain the honor of what it is that they serve. Honor is the most important attribute of the warrior. Integrity is the bedrock of honor. *Honor is the highest virtue of warriorship.* Without honor, a person is not, and cannot be, a warrior.

> "You may abandon your own body but you must preserve your honor."
>
> – Myamoto Musashi (Japanese swordsman and strategist)

There is nothing easy about living up to any of these attributes. The warrior takes them as a personal challenge for self-improvement.

> A Soldier's Prayer:
>
> "Lord, let me not prove unworthy of my brothers."

The origin of this prayer is unknown. It is sometimes used by U.S. military personnel. Note that the prayer does not ask for personal protection.

CHAPTER THREE

THE QUALITIES OF THE WARRIOR

Contained within these attributes are four qualities which must also be discussed. These qualities are initiative, self-reliance, discipline, and resiliency.

INITIATIVE

Warriors are capable of independent action and are prepared to act. If they see something that needs to be done, they do it. While the warrior is part of a team and is a good teammate, the success of the team also depends upon the warrior seizing the initiative when necessary.

SELF-RELIANCE

The warrior is confident to be self-reliant when necessary. By not letting him or herself down, he or she does not let the team down. Through self-knowledge, the warrior will know when it is necessary to seize the initiative.

Hollywood would have us believe that the hallmarks of a warrior are initiative and self-reliance – independent action, like the lone wolf operating on his own. Indeed, a warrior must be *capable* of these things, but is foremost a team member.

It is true that there are figures from history who are noteworthy for their exploits in single combat. They have accomplished great achievements and are celebrated as icons of independent action. But what makes these lone wolf figures of history – and those that appear in fiction – so notable and compelling is that *they are unusual*. Most warriors, most of the time, fight within teams.

Alexander the Great is celebrated as a great king and general, one whose prowess in battle was legendary, leading from the front. We focus on Alexander, the great figure from history, but in doing so we forget that he was a member of an army. He was both the leader of and a member of a team. Alexander never forgot this, and neither should we. Even the leader is a part of the team. And a leader cannot lead if there is no one to follow them.

Independent, lone wolf action is undertaken when necessity requires, and a good warrior is capable of it. There are individual moments where lone action and bravery come into play. But these individual actions are conducted as part of a team. Individual heroism should not be discounted, but we must be mindful of the greater context of teamwork.

A warrior puts the team first. Life is a team sport, after all. Remember, "Rambo" is fiction.[7]

> "An army is a team. It lives, eats, sleeps, and fights as a team. This individual hero stuff is bullshit."[8]
>
> — General George S. Patton, Jr., speech to the U.S. 3rd Army, World War II

DISCIPLINE

The warrior is disciplined. This means holding oneself accountable and taking corrective action. It also means being

---

[7] "Rambo" is a character in a five-part movie series spanning nearly forty years from 1982 to 2019, played by actor Sylvester Stallone. The movies are based on the 1972 book, *First Blood*, by David Morrell.

[8] General George S. Patton, Jr., speech to the U.S. Third Army, from "*Patton, Montgomery, Rommel: Masters of War*," 2009.

a disciplined teammate and team member. The warrior's discipline is directly tied to the warrior's attributes.

The attributes of the warrior we discussed in Chapter Two require discipline. Without exercising self-discipline, warriors will fail to live up to the attributes they aspire to uphold.

"Through discipline comes freedom."

— Aristotle (Ancient Greek philosopher)

With discipline, one is free to do more, to do other things. Think of it this way: if a person lacks discipline and sits around watching videos all day when he really should be working, working out, doing chores, or whatever needs to be done in order to actually accomplish something, he is a slave to sitting around watching videos.

Those things that need to get done just pile up. They must still get done, but the workload has actually increased for the person because now he has to get much more done in a shorter period of time.

If he were disciplined, he would do all the things that need to be done *and* have time left over. He would have additional time left over for sitting around and watching videos or whatever other recreation he prefers. In fact, his options to do more things increase because his discipline has given him more time. And time is the one thing that we cannot get more of – ever.

RESILIENCE

Resilience is the ability to adapt to novel situations and to recover quickly from setbacks. The warrior is both physically and mentally resilient. Physical resilience supports mental and emotional resilience. Physical exercise, the playing of sports, the study of martial arts, and physical competition of all kinds play a role in developing and maintaining physical resilience.

Physical labor and outdoor activities – particularly hunting, fishing, and camping – also help develop and maintain resilience. Challenging the body also challenges the mind. A hard workout may cause one to want to quit, and it is that "nope, I'm finishing this set," or "nope, I'm getting one more rep," attitude that helps create a tougher-minded person. There are other ways to do this, but exercise is the easiest way to start toughening the mind.

Seeking discomfort by doing activities outside of one's comfort zone is key to building resilience. It helps enable the warrior to be comfortable in uncomfortable situations. Being comfortable in uncomfortable situations is a form of resilience.

Emotional resilience must be built, too. It can be developed through meeting challenges. The activities suggested above can certainly help with this, especially doing things that are squarely outside one's comfort zone.

One of my closest friends with whom I grew up spent many, many years as a youth sports coach. He has coached football, baseball, softball, soccer, and wrestling. He is also an experienced outdoor educator, having spent years teaching youth about the great outdoors, fieldcraft, and other outdoor activities, including challenge courses and rope confidence courses.[9]

Several years ago, he was the head of outdoor education at a large church-based summer camp. His students came from all sorts of backgrounds and included kids from very wealthy families. He observed that the kids from wealthy families appeared to have a harder time participating in and performing various activities, particularly the challenge courses.

In running kids through challenge and rope courses, he noticed that the kids from well-off families were easily frustrated when attempting to negotiate obstacles on the courses. In contrast, though, kids from underprivileged and working-class homes seemed to do fine, even to thrive, while negotiating these obstacles.

---

[9] An outdoor educator teaches outdoor activities to children who do not usually have the opportunity to engage in them. A challenge course is similar to an obstacle course and seeks to challenge decision-making and team-effort skills against the backdrop of negotiating various obstacles.

Time and time again he noticed that a child's ability to negotiate the challenge and rope courses seemed somehow correlated to their socio-economic situation. The lower they were on the socio-economic ladder, the better they were able to negotiate the courses. The higher on this ladder, the harder it was for the kids.

He wanted to understand why his observations kept being repeated over the course of successive summers and with different groups of different people rotating through the encampment as these summers went along.

The coach took time to talk with the parents when he had the chance, to try to understand the home lives these kids were leading. What he gleaned from his many conversations was that the more the parents coddled their kids, the less capable the kids were on these courses. In contrast, kids who did not have it easy at home, who had to make do with what they had and who had to do things for themselves, had a much easier time on the courses.

The coach related that he does not believe the difference is strictly economic, but rather that it is a matter of the children failing to develop resiliency skills. He observed that students from disadvantaged backgrounds have more resilience out of necessity. Conversely, a great number of affluent parents tended to remove too many obstacles from their children's lives, which resulted in a lower level of resiliency. He also observed this effect in many middle-class students.

What I am relating to you is anecdotal, not empirical, for sure. But it is indicative of the ease of our society. The coach says that he "can tell how well or poorly a kid will do on a challenge course by knowing where the kid is from."

The sociologists and psychologists can study the "root causes" of this type of behavior. This book is not an academic report. But there is a lesson here, at least a layman's lesson, in what modern society with its relative ease has created. This ease offers little opportunity to grow through the meeting of challenges.

The warrior must be willing and able to face challenges and learn from them. Through the facing of challenges comes personal growth and greater resilience. Through resilience comes self-reliance.

CHAPTER FOUR

WARRIORS LIVE BY A CODE

The warrior's code, also known as a credo or ethos, binds the ancient warrior to the modern warrior.[10] The *way* wars are fought changes over time. For example, the Roman legions did not have fighter aircraft flying cover for them. But the *type of person* who fights wars – the warrior – does not change. Wars change; warriors don't.

Historically, there has been no single accepted warrior code. Each society, culture, time, and place and the necessities of the day have rendered different codes over time. But what is striking are the similarities between different warrior codes over time, place, and culture.

Beowulf is an Old English epic written around 1,000 AD/CE. The warrior code offered in this work is this: strength, courage,

---

[10] "Warrior ethos," "warrior code," and "core values" are used interchangeably in this book.

generosity, and honor.[11] Moving forward a few hundred years to the Medieval period, the Middle Ages, we have the code of the chivalric warrior.[12]

While there was no one accepted warrior's code during the Middle Ages, from the Duke of Burgundy in the 14th Century, we can list the following attributes of a chivalric warrior: faith, charity, justice, sagacity, prudence, temperance, resolution, truth, liberality, diligence, hope, and valor.[13]

Compare these to the Four Virtues of the Lakota Sioux:[14] courage, generosity, wisdom, and respect. Can you see the similarities? Valor = courage; charity = generosity; sagacity = wisdom; liberality = respect. When comparing the Four Virtues of the Lakota Sioux to the code presented in Beowulf, the similarities are noteworthy!

---

[11] No distinct listing like this appears in the poem. This list is derived from the overall text and is delivered here in Modern English.

[12] Chivalry: the system of behavior followed by knights in the Medieval Period of history that put a high value on honor, kindness, and courage. Cambridge English Dictionary, 2022.

[13] *The Telegraph*, London, U.K., "Chivalry explained: from knights of honour to women's lib," Emily Gosden, June 15, 2011.

[14] *Lakota Life*, 1986.

The Lakota Sioux did not read Beowulf, and the Anglo-Saxons did not collaborate across the centuries and across the continents with the Lakota Sioux. These codes are recognized by the human warrior across time and place. Arguably, the warrior code is universal, even if it is expressed using different words.

The "Ways of the African Warrior"[15] included being a skillful fighter and hunter, being noble and respectful, being knowledgeable of the healing arts, being in harmony with nature, and having faith in God. Following these ways, the African warrior could fight to defend the tribe, hunt to feed the tribe, behave with nobility and be respectful of others, and could treat wounds received in battle or while hunting. He believed he would also have the support that Mother Earth and faith in a higher power can provide.

The Japanese code of Bushido is an oft-cited warrior code. The virtues of Bushido are righteousness, heroic courage,

---

[15] *Shiai Magazine*, Republic of Cameroon, "Seven Ways of the African Warrior," Aurelien Henry Obama, April 4, 2008.

benevolence and compassion, respect, honesty, honor, loyalty and duty, and self-control.[16]

We can also look to the core values of the U.S. Marine Corps: honor, courage, commitment. The core values of the Federal Bureau of Investigation are fidelity, bravery, and integrity. Again, the similarities are obvious and striking.

War is timeless, warriors are timeless, and the warrior lives by a code! While the means of war changes over time – the tools, weapons, technology may change – the nature of war and of the warrior do not. What binds the ancient warrior to the modern warrior, what binds all warriors across all time is the warrior code!

> "War is timeless and ever changing. While the basic nature of war is constant, the means and methods we use evolve continuously."
>
> – *Fleet Marine Force Manual 1, Warfighting.* U.S. Marine Corps

---

[16] From *Bushido, the Soul of Japan*, 1900.

CHAPTER FIVE

WHAT IS YOUR CODE?

It is important that the warrior has a personal code. It is central to the very being of a warrior. It is the warrior's moral, ethical, and martial compass. Without a personal code, the warrior is missing an essential touchstone of warriorship – an answer to the question of "What guides me as a warrior?" A warrior code is a statement of the *core values* held by that warrior, and of the warrior group to which the individual warrior belongs or aspires to belong.

There are many sources for a warrior code. The U.S. military services, native tribal warrior codes, ancient Anglo-Saxon codes, ancient Nordic codes, or those adopted by first-responder agencies are but some examples of where you can find your warrior code, your core values which you adopt, or accept, as a warrior.

It is important to recap what we discussed in Chapter Two, under the heading of "Selflessness:" the warrior is not seditious, treasonous, or criminal. The warrior is honorable!

For example, I have adopted the core values of the United States Air Force as my warrior code. These core values are *"Integrity First, Service Before Self, Excellence in All We Do."* It may seem strange that a former U.S. Marine has chosen these core values as his warrior code, but as of the time of writing, I am a nearly sixty-year-old husband, dad, small business owner, retired lawman, retired soldier, and a former elected official. This warrior code fits my life best.[17]

Let's take these core values one-by-one. *Integrity First!* For me, this says it all! What are we without integrity? Integrity – doing what is right because it is the right thing to do, to be honest even when no one is looking – is honorable. Integrity is the bedrock of honor!

Service before self is another foundational principle. Serving something or someone before your own needs is a

---

[17] We may change our perspective, thus our code, at different stages of our lives. There is nothing wrong with this, nor is it wrong to adhere to the same code throughout one's lifetime. This is an individual decision that is based upon that individual's circumstances. This is part of our growth as human beings.

requirement of both leadership and of teamwork. Do not let your brothers in arms down and leave no man behind.

Excellence in all we do is a guidepost to always do one's best. And, in those times when we don't, we should resolve to do better next time.

I very strongly recommend that you adopt your own core values, your own warrior code, and then strive to live by that code each day. This is necessary, in my view, to live like a warrior. It will help you be a better person and a better warrior!

CHAPTER SIX

THE INNER BATTLE

Everyone is fighting for something. Perhaps someone is fighting for their life in battle against cancer, or maybe someone is fighting for the survival of their marriage. Perhaps someone is fighting to pass math class or to break a bad habit, or just for a happy life.

Much of the time, these fights, these battles, are happening on the inside of the person, where we cannot see it. This is the inner battle, and we all have at least one that we are fighting all the time. Some people have more than one. All of us are fighting for something, *all the time*.

The inner battle is the fight that is the most difficult. It is within ourselves and, quite often, it is *with* ourselves. The measure of the warrior is how the warrior fights these inner battles, and this is where strength really matters. Having a warrior code is vital to facing the inner battle. It is one source of strength, along with the warrior's faith system, family, and friendships.[18]

---

[18] Please see the Note on Faith on page 79.

The inner battle is often the most devastating battle that the warrior ever fights. Especially when we are our own opponent! Many warriors fail at the inner battle, and concentrate only on external fights, the outer battle. Or, to put it another way, many warriors *want to **only*** concentrate on the external fight. In many ways, that fight is easier.

It is of the utmost importance that you, the warrior, be aware of this and prepare yourself for this fight over any other fight.

One of the most significant Achilles' heels of the warrior is guilt. Guilt is not an altogether bad thing; it is a product of our conscience. It is the little voice inside us that tells us, "You messed up and you need to make it right." But sometimes, especially during or after a war or other periods of hyper stress, guilt can cripple us. It could be "survivor's guilt," feeling that "I should have done more," or "I could have saved him/her," "If I had only done _____, then everything would be better, everything would be all right." Maybe. But on the other hand, probably not.

"In war, you can do everything right and still die."

- Brigadier General James M. Stewart, USAFR (Ret.), (B-24 pilot, squadron commander, and group commander, European Theater, World War II[19])

Warriors are task-oriented, practical people. When the accomplishment of a task, a goal, or an objective fails, warriors often burden themselves with the failure, even if it was through no fault of their own. It is part of the warrior culture, the warrior mindset. It must be, and here lies the difficulty. "Failure is not an option" is true; failure can never be an *option*. A warrior does not start out *conceding* failure. But failure does happen. It is an unhappy condition of life. Learning to deal with failure is vital to surviving it. Warriors are only human.

Warriors are taught that getting knocked down is not the measure of the warrior's mettle. The measure is whether or not the warrior gets back up. When faced with something that

---

[19] Recipient of two Distinguished Flying Crosses and four Air Medals for combat, among many other military awards. Also known as Jimmy Stewart, Academy Award winning actor. This quote was attributed to him in a publication long forgotten by your author. It has likely been uttered by many soldiers throughout the millennia.

knocks them down, they get back up. But sometimes warriors cannot understand why, or how, they were knocked down in the first place. "This shouldn't have happened to me, that way," is a typical thought. "I should have been better! Why wasn't I better?"

These thoughts fuel uncertainty, which can become an obsession, especially when the warrior believes that he or she could have, singlehandedly, changed the outcome of a particular situation. The warrior comes to question his or her own character and capabilities.

We see it in warriors who have been in combat. An individual might hold a deep-seated conviction that if only they had put more rounds on that upper-left-corner window where the insurgent was firing from, or shouted "incoming!" a second sooner, a fellow warrior would be alive, or unwounded, and that the day would have turned out differently. Maybe. But probably not. "In war, you can do everything right, and everything can still go wrong," to paraphrase the previous quote. Warriors are only human. The superhuman warrior is found only in comic books.

Each warrior must understand this about themselves. The only thing we can ask of ourselves is to do our best, and when we don't, to resolve to do better. It is easy to say, but hard to do, even though the concept itself is simple. Another quotation is perhaps helpful in this part of our conversation:

> "Everything in war is simple, but the simplest thing is difficult."
>
> – Carl von Clausewitz[20]

This can be said of many things in life.

Warriors are susceptible to feelings of guilt over their performance or their survival, and this is unhealthy. Therefore, the key to winning the inner battle is for you to understand that you are only human, and that forgiving yourself – and allowing yourself to move on – is the way forward to the next objective.

---

[20] Carl von Clausewitz was a general in the Prussian Army during the first half of the 19th Century. He is a noted thinker on war and military affairs. From *On War*.

This is so even if you believe you do not *deserve* to move forward. This is a version of picking oneself up after being knocked down and getting on with things. The warrior *does deserve* to move forward; *YOU* deserve to move forward, because you owe it to other warriors, past, present, and future to do so. After all, who will show the way to other warriors, and to warriors on the way, than **you**?

> "If the mind is to emerge unscathed from this relentless struggle with the unforeseen, two qualities are indispensable: first, an intellect that, even in darkest hour, retains some glimmerings of the inner light which leads to truth; and second, the courage to follow this faint light wherever it may lead."
>
> — Carl von Clausewitz[21]

The mind will not always emerge unscathed. But it is that inner light that burns within every warrior – that warrior ethos or warrior code – along with whichever spiritual faith the warrior possesses, that allows the warrior to shake off guilt – and doubt – and regain his or her compass bearing and march onward.

---

[21] Ibid.

Even if the warrior is the most perfect, greatest, and most talented warrior in the entire history of humanity, things would still go wrong. When warriors judge themselves against *perfection*, they will *always* fall short. You can strive for perfection, but you must realize – and admit – that perfection eludes each and every human being.[22]

Guilt does not have to be the only thing that hurts the warrior. Feelings of inadequacy can do nearly as much damage. So does "imposter syndrome."[23] These are but examples, and you may be fighting an inner battle with something else. If so, remember your touchstone, your core values, your warrior code.

I want to be clear with you here: having a warrior code or adhering to a set of core values is not a cure-all – not by a

---

[22] A paraphrase from the novel "*The Lost Fleet – Guardian*," by Jack Campbell, 2013. Thank you to John G. Hemry.

[23] Imposter syndrome is not an official diagnosis recognized by the American Psychological Association. However, it is the "condition" in which a person does not believe that they are worthy of whatever accomplishments they have achieved, and/or are not worthy of any accolades.

longshot – but it is a big help. And often it is just enough help, along with faith, family, and friends, that we warriors need to enable us to go on about our business, to go on with our lives. Live your code!

We owe it to those warriors who have come before us, and to those who will come after us, to set the example and move forward in life. Be the example, whether you think you should be or not. I believe that you should be.

CHAPTER SEVEN

THE OUTER BATTLE

Generally, the outer battle is easier to see and easier to fight. An example of an outer battle is an automobile accident. It happened, your car or truck is totaled and now you must deal with it. You have to deal with your insurance company, the other driver's insurance company, the police, doctors, and hospitals if you or someone else was injured, etcetera. You have this "battle" to fight, and everyone can see it. It is obvious.

The reason we talk about the outer battle – and everyone has one of some sort going on at any given time – is that it is the opposite side of the coin of the inner battle. The outer battle can cause an inner battle, or it can mask an inner battle.

It is easier for warriors to concentrate on fighting the outer battle, to "take care of business." Resolving the outer battle provides the warrior with a tangible victory that others can see. Warriors are "get it done" types of people. Fighting the outer battle and overcoming the obstacles contained in that battle provides satisfaction to the warrior.

The outer battle could be one's battle in recovering from illness, wounds, or injury,[24] such as having lost a limb in combat. Everyone sees that you are engaged in this battle, which makes it different from the hidden, inner battle. But the two are linked, which is why this chapter is in this book.

You see, being a warrior is not without consequence. There may be a price to pay, and some warriors have paid an enormous price. Mental, moral, and physical injury are dangers faced. So is death. All warriors need to be aware of these things, because through awareness, they can develop strategies to deal with these battles when they arise. They can be mentally prepared for them, even if only through expectation. Through this expectation comes preparedness.

In some ways, life is about facing one problem and then the next one, and so on. This is a constant, along with the certainty

---

[24] For the purposes of our conversation, a wound is intentionally inflicted; an injury is the result of accident.

of death and taxes. There is always another problem to resolve, another issue to handle. On top of these things is a contradiction: change is always afoot. It seems like nothing changes and yet everything changes, all the time. Change is constant; expect these changes and thus be prepared for them.

Challenges and change can be overwhelming, to be sure. This is the chaos of life, and it is not unique to you. Life is tough, and everyone experiences this chaos, whether you see it in others or not. No one goes through life unscathed – *no one*.

In the midst of these unchanging constants and in the midst of this constant change, you have your inner and outer battles. It is important to understand that this fact is not unique, even if the specific challenges or difficulties are unique.

If you are engaged in an outer battle, do not let it prevent you from also fighting your inner battle. Also, do not let it mask your inner battle. You can fight both at the same time. The inner battle is more difficult, so it is easier to focus solely on the outer battle. But you can do both. You can do both in the midst of the chaos of life.

Remember, too, that the warrior is part of a team. Life is a team sport. It is not a weakness to ask for help when everything going on is overwhelming, when it becomes "too much to handle." The warrior is only human after all. The superhuman warrior exists only in comic books. Ask your fellow warriors for help.

In none of the warrior attributes nor any the warrior codes that we have discussed is it stated or implied that asking for help is a weakness. It is not. If you need help, ask a teammate. If a teammate asks for help, provide it. Good teammates do this. Do not think that you must go at it alone; turn to your fellow warriors. And, as always, adhere to your warrior code.

The only one who can stop you is you. No one said these things are easy, but they are not impossible. You can succeed.

"Living successfully is not nearly as difficult as living unsuccessfully."

— Unknown sage

CHAPTER EIGHT

WHEN THE WARRIOR IS NOT A SOLDIER

What if you never were, are not, and never will be a soldier, or anything related to war?[25] Can you be a warrior? The answer is that you can certainly live like a warrior, walk your own Warrior Path, and become the warrior that is within you.

In order to be a true warrior, you must live according to a set of warrior core values. You must also live according to the attributes and qualities of the warrior as we discussed in Chapters Two and Three. And importantly, *a true warrior serves something greater than themselves.*

While honor is the highest virtue of warriorship, it is *service* that distinguishes the warrior from the non-warrior. The warrior serves something greater than themselves. This is the most quintessential aspect of living like, and being, a warrior: *service before self!*

---

[25] Soldier is being used holistically here. Marine, soldier, sailor, airman, guardian, and sometimes emergency services personnel are encapsulated in the word "soldier." (Coast Guard personnel are encapsulated in the word "sailor.")

There are certain things you can do to be a warrior, to become a warrior, to live as a warrior. The first step is to adopt a warrior code.[26]

Search for the warrior code that best fits you and your life situation. You can refer to Chapter Four for some ideas. You can search the internet or a library for a warrior code that speaks to you the best.

However, creating your own code is not recommended. When we create our own, we tend to be very self-serving. Rather than create something to aspire to, we create something that is easy, which allows us to excuse ourselves from our occasional, less-than-ideal, behavior.

There are plenty of warrior societies that have developed their own codes. You do not need to create your own. Find one that fits best for you and live by that code! Think of this as following your warrior path.

---

[26] These steps are arbitrary and are the products of my advice given to you as part of our conversation.

You may fall short sometimes, but so what? So does everybody. Just do better next time. To use a phrase I've heard a time or two, "Shut up, ruck up, and drive on."[27]

The second step is to find something bigger than yourself to serve. *This is the vital component of warriorship!* A warrior does not just serve himself. A warrior who only serves himself is not a warrior at all. Not in any way, shape, or form. *A warrior serves others!*

Adopt the spirit of volunteerism. You can serve a nation, a state, a community, an ideal, or a cause. Such service can be professional, volunteer, or charitable. There are many ways to serve others.

There are many opportunities to volunteer for the greater good. Become a volunteer firefighter and/or an emergency medical technician. Apply to be a reserve or auxiliary police officer in your community.

---

[27] "Shut up" means quit griping. "Ruck up" means to put your pack on or get on with your work. "Drive on" means to get moving. Put another way, "Quit bitchin,' get your shit together, and get going." I've heard that a time or two, too.

Many cities in the United States have a Citizens on Patrol organization within their police departments. Citizens on Patrol, or COPS for short, are unarmed civilian volunteers who augment police departments as a sort of neighborhood watch on steroids.

Both the Civil Air Patrol as the Auxiliary of the U.S. Air Force, and the U.S. Coast Guard Auxiliary offer civilian volunteer opportunities which can be very rewarding and fun. Both organizations assist in search and rescue and are great service organizations. You do not have to meet the qualifications for military service to join either of these organizations.

There are more volunteer opportunities than most people realize, and too many to list here. The examples above are just a small selection. Seek these opportunities out, and when you find one that suits you, volunteer and serve.

The third step is to work on your personal fitness. This means mental fitness, spiritual fitness, and of course, physical fitness. The study of martial arts is a great way to improve your personal fitness. There are many types and styles to choose from, so you could find one that fits you and begin your martial arts journey. Exercise in some form – find what you like to do, go do it, and keep doing it.

The fourth step is to learn a skill that is outside your current skill set. Learn something new. This new skill should be useful to both yourself and to others. Examples of useful skills to learn are carpentry, mechanical repair, plumbing, and electrical work.

If you do not already do so, learn to hunt and fish.[28] Learn to grow your own fruits and vegetables. These skills will help you feed yourself and can help you feed others.

Cooking provides useful skills worth learning. If you do not know how, learn to cook, especially over an open fire and outside of a well-equipped kitchen. And I do not mean just the backyard grill. Such skills are very practical and can serve you as well as others.

---

[28] Hunting and fishing ethics demand that you, or someone else, benefit from what you take. Game should be harvested as a food source, not only as a trophy.

Explore a new interest, and if it suits you, learn the ins and outs of it. Enjoy the learning, and that enjoyment will lead to knowledge. The warrior not only serves others but is constantly growing as a person. Developing new skills and learning are pathways of growth. Always learn and grow.[29]

> "Without knowledge, skill cannot be focused. Without skill, strength cannot be brought to bear and without strength, knowledge may not be applied."
>
> — Alexander the Great

The fifth step is to read about warriors, warriorship, and warrior skills. We live in an era of online videos. There is nothing wrong with videos, but the study of a thing requires, at least in my view, reading and reflection upon what is being read. Reading is essential to developing your warriorship. Of course, watch the videos, too.

---

[29] See the Note on Learning Self-Sufficiency on page 82. Inspired by author Varg Freeborn.

On reading:

> "The problem with being too busy to read is that you learn by experience (or by your men's experience), i.e., the hard way. By reading, you learn through others' experiences, generally a better way to do business, especially in our line of work where the consequences of incompetence are so final for young men. Thanks to my reading, I have never been caught flat-footed by any situation, never at a loss for how any problem has been addressed (successfully or unsuccessfully) before. It doesn't give me all the answers, but it lights what is often a dark path ahead."[30]

>   — General James N. Mattis, USMC
>   26th U.S. Secretary of Defense

You do not have to be a general to read. All warriors, whether in uniform or not, should read about warriorship, and about their profession or vocation. Reading expands one's horizons, perspectives, and thinking. Readers are leaders, and warriors are leaders.[31] Please see the section on Recommended Reading on page 73.

---

[30] From "*Call Sign Chaos*," Appendix B, Page 256.
[31] Please see the Note on Reading for U.S. Marines on page 83.

On sports:

If you are young enough to do so, play games that develop teamwork and offer adversity. Sports such as gridiron football, rugby, lacrosse, ice hockey, Gaelic football, Australian Rules Football, water polo, and field hockey are some good options. These are collision sports and offer tremendous lessons in overcoming adversity.

The martial arts are also excellent in this regard. Boxing, wrestling, judo, Brazilian jiu jitsu, Muay Thai, and the "traditional" martial arts that offer live resistance training are excellent.

For the development of athletic skill outside the realm of the collision sports are baseball, softball, basketball, and soccer, all of which are great for kids, allowing them to develop teamwork skills and athletic ability. Soccer, baseball, softball, and basketball can be great precursors to entering collision sports.[32]

Baseball and softball are particularly excellent games in having both individual and team elements. When you are on offense (at bat), it is like an individual sport. It is just you and the pitcher. When you are on defense, it is a team effort. It's the best of both worlds.

There is nothing wrong with individual sports such as track and field, golf, tennis, etcetera. These are all beneficial in multiple ways for fitness, developing a competitive edge, and overcoming challenges, to name a few. But I strongly encourage kids to play at least one team sport to develop teamwork skills. Remember, life is a team sport.

If you are past your athletic prime, exercising for physical fitness is recommended; in fact, it is necessary in my view. Lift

---

[32] Soccer and basketball are contact sports. They are not, however, collision sports. That is not to say that collisions do not occur, only that colliding is not a requirement to advance the game.

weights, run or hike or walk, ride a bicycle, perform calisthenics, yoga, and the like, to maintain your vitality.

You can find the fitness modality or modalities that work best for you. Exercise is not only a requirement of warriorship, but it is necessary to maintain your warrior vitality.

Do not let your age discourage you from the study of the martial arts or from exercise. Older warriors must train smarter, not harder, but they can still train. And warriors do train, especially the grey-beards. The grey-bearded warrior must train to maintain his warriorship. Get out there and train!

CHAPTER NINE

HONOR

The highest warrior virtue is honor. A warrior without honor is no warrior at all. A warrior without honor is simply a thug.[33]

The warrior strictly adheres to the rules – the laws – set forth by his or her society. When his society establishes additional regulations that govern the warrior's behavior, whether those are Rules of Engagement or Use of Force policies, the warrior is strictly obedient to them.[34]

At the same time, the warrior's honor is not bestowed by society and its rules. A warrior's honor comes from within, and that honor is *sacred* to the warrior.

---

[33] A "thug" is someone who is violent, especially in committing crimes. Cambridge English Dictionary, 2022. In modern American society, the word "thug" has multiple meanings, and has even taken on the aura of a compliment. While your author believes this to be an assault on the English language (pun intended), as used here it means a violent, self-serving criminal.

[34] This does not mean to say that a rational, reasoned departure from these constraints cannot be made based upon unusual, extreme, or unanticipated circumstances. However, such a departure must be justifiable.

Warriors expect fellow warriors to uphold not only their own honor, but that of their fellows. In other words, warriors do not tolerate behavior from their fellows which besmirches their own honor. In the U.S. Marine Corps, honor was often enforced by the use of a simple admonition, from one Marine to another: "Marines don't do that."

Warriors do not use their skills or power to oppress the weak, to take advantage of the vulnerable, or to take from others. On the contrary, the warrior defends the weak, protects the vulnerable, and respects persons and their property. The warrior is a defender and a protector and is an enforcer only when it is necessary. Warriors destroy only when it is the only reasonable, rational option remaining.

The warrior acts with reason and restraint, not barbarous mayhem. The warrior has purpose which serves something greater, and is not motivated by self-interest, but rather is motivated by service to others.

If warriors are called upon to use force, they do not do so indiscriminately. They use only the force necessary to accomplish the objective and use that force only against those who are foes, safeguarding those who are not.[35]

We honor our foes by extending to them basic human dignity and respecting their human rights. When it is most difficult and distasteful to do so, extending this dignity *elevates* the warrior's own honor.

The Honor Code of the U.S. Military Academy at West Point[36] states that, "*I will not lie, cheat, steal, or tolerate those who do.*" This code is simple and straightforward, and one which all warriors – whether military or not – would be in good stead to follow. This is a simple formula to guide the honorable warrior along his path of honor.

---

[35] It may be that a stern admonition, or one punch in the nose, or one 7.62mm round, or 100 two-thousand pound bombs is that force which is necessary. It will depend on the circumstances.
[36] The Honor Code is not a warrior code, but it is closely aligned with warrior codes.

And it is just plain good advice for anyone, anywhere. Of course, always remember and live by your warrior code.

CONCLUSION

I believe that it is clear that warriors are something special – not unique, but special. There are people who have jobs that require them to run *toward* the sound of gunfire, not away from it, run *into* burning buildings to save people, not run from them. There are people who choose to be protectors, defenders, sometimes enforcers, and when absolutely necessary, destroyers, *in the protection of others.*[37] These are the warriors.

Some people choose to forego lucrative and prestigious jobs to toil away in a medical laboratory seeking cures for some of the toughest diseases that befall humanity. Others choose lives in difficult places under difficult circumstances to bring fresh water to people who do not have it, to help starving people raise their own crops and grow their own food.

There are those who choose to teach the young rather than embark upon much more highly paid careers. There are those who choose to wear a uniform which may require them to go

---

[37] Please see the Note on the Nature of the Warrior and War on page 84.

into harm's way and make the ultimate sacrifice – a sacrifice of themselves for others. These are the warriors.

Human beings have an innate need to have purpose, to have something to live for, to have something that gives meaning to their lives. Man's search for meaning is universal, and it is no different with the warrior. Indeed, it is elevated with the warrior.

For the warrior, meaning is found in purpose and the warrior's purpose is service. In military parlance, the mission is the warrior's essential purpose, and that mission is to serve something greater than the warrior's self.

It is this service that distinguishes the warrior from the non-warrior. While I am certainly biased toward the military warrior, a person need not be a soldier, need not wear a uniform, to live like a warrior.

Warriors' service to something greater than themselves must also come with some cost to themselves, whether this cost is in comfort, money, prestige, or in some cases, their life. And this service must be done with honor because the warrior is, above all, honorable. Honor is the highest virtue of warriorship.

> "Don't point fingers, don't complain, get out there and do something. Service before self!"
>
> — Major Jason Allred

Honor is based upon integrity and the warrior's honor comes from within; it is not imposed from outside. Honor may be expected of the warrior by something, or someone, outside of the warrior, but being honorable is an inner trait, an inner resolve, not something someone is ordered to have.

Warriors honors their society, their fellows, their purpose, their work, and even their foes. The warrior seeks to never stain his or her honor, and to never stain the honor of what it is that he or she serves.

A warrior code is the statement of the core values held by the warrior. It is the guidepost by which the warrior upholds his or her honor and guides his or her service to others.

Therefore, if you wish to be a warrior, you must choose a warrior code. This will serve as your ethical, moral, and martial compass. Without a personal code, you are missing an essential touchstone of warriorship – an answer to the question of "What guides me as a warrior?"

What binds the ancient warrior to the modern warrior, what binds all warriors across all time is the warrior's code! This is true whether you are a cancer warrior fighting disease or a special operations warrior fighting your country's enemies. Adopt your own warrior code and make it the codex of your life.

Now, go forth on your warrior path.

STANDING FAST

In times of calm, there are those who stand fast on the ramparts.

When challenges come, there are those who stand fast at the bulwarks.

If peace shatters, those who stand fast sally forth into the breach.

When those who sallied forth may fall, there are those standing fast to take their places.

In victory, there are those who stand fast to remember the fallen.

And again in peace, there are those who stand fast, manning the ramparts, to ensure that peace prevails.

It is the warrior who stands fast.

By Todd Kenneth Hulsey
© 2019

Comanche brave – a warrior of the Southwestern U.S. plains tribe known as the Comanche.

Hunkpapa Lakota – a Native American tribe of the Great Plains. Also known as the Lakota Sioux.

Martial arts – "Martial", derived from the name of the Roman god of war, Mars, means military or fighting. "Arts" is derived from artisan, or a craftsman, and can be translated as "skill." Martial arts can be thus defined as "fighting skill." There is also the physical art of human bodily movement displayed when practicing or using martial arts.

MMA – Mixed martial arts – A competitive fighting sport in which two contestants use a mix of different types of martial arts. The most typical martial arts seen in MMA are boxing, kickboxing, Muay Thai, wrestling, Brazilian jiu jitsu, and judo. Other martial arts such as sambo and karate are also seen in MMA.

Roman legionary – A soldier in the army of Rome. Rome, an Italian city-state that went on to rule all of the Mediterranean and most of Western Europe, was founded in 509 B.C./B.C.E. It was a republic for about 500 years and an empire for roughly the next 500 years. The western half of the empire fell in the fifth century A.D./C.E., while the eastern half continued until overtaken by the Ottoman Empire in 1453 A.D./C.E. The Eastern Roman Empire was also called the Byzantine Empire. The Roman legionary was the best trained, best equipped, most fit, and best led soldier in the world for arguably over

400 years from roughly 125 B.C./B.C.E. to 275 A.D./C.E.; some would say for a longer period.

Warrior – A person trained for and prepared for engaging in that enduring, quintessential human activity that we call war; one who is intellectually, emotionally, and spiritually ready to engage in warfare in all its forms but most particularly combat on the battlefield, whether it be on land, sea or in the air or space; and who is capable of taking human life and sacrificing their own in service to something greater than themselves. It is also someone not trained for war but who is honorable, follows a warrior code, and serves something greater than him or herself, and seeks to serve others. A warrior is a protector, a defender, sometimes and enforcer, and when all else fails, is sometimes a destroyer in the protection of others.

Warrior code – A statement of the core values held by the warrior; the warrior's guidepost which guides him or her on their warrior path.

Warriorship – the state or condition of being a warrior.

BIBLIOGRAPHY

Brighton, Terry, *Patton, Montgomery, Rommel: Masters of War*, New York, N.Y., Crown Publishing Group, 2009

Campbell, Jack, *The Lost Fleet – Guardian*, New York, N.Y., Penguin Group, LLC, 2013; LtCdr John G. Hemry writing as Jack Campbell

Clausewitz, MajGen Carl (Carl von Clausewitz), *On War*, Princeton, N.J., Princeton University Press, 1989

Cotton Vitellius A xv., *Beowulf, A Student's Edition*, translation by E.L. Risden, Troy, N.Y., The Whitson Publishing Company, 1994

Curry, Oliver S., Mullins, Daniel A., Whitehouse, Harvey, *"Is It Good to Cooperate? Testing the Theory of Morality-as-Cooperation in 60 Societies,"* Chicago, IL., Current Anthropology, University of Chicago Press, Volume 60, Number 1, February 2019

French, Shannon E., *The Code of the Warrior*, Lanham, MD., Rowman & Littlefield Publishers, Inc., 2003

Freeborn, Varg, *Violence of Mind*, Warren, OH., One Life Defense, LLC, 2018

Gosden, Emily, *Chivalry explained: from knight's honour to women's lib*, London, U.K., The Telegraph, June 15, 2011

Kaufman, Stephen F., *The Martial Artist's Book of Five Rings*, Boston, MA., Charles E. Tuttle Publishing, 1994

Krulak, LtGen Victor H., *First to Fight, An Inside View of the U.S. Marine Corps*, Annapolis, MD., Naval Institute Press, 1984

Mattis, Gen James N. & West, Francis J., *Call Sign Chaos*, New York, N.Y., Random House, 2019

Morgan, LtCol Forrest E., *Living the Martial Way*, New York, N.Y., Barricade Books, 1992

Nitobe, Inazo, *Bushido, The Soul of Japan*, Oviedo-Asturias, Spain, King Solomon Books, 2020. Originally published in 1900 by The Leeds & Biddle Company, Philadelphia, PA

Obama, Aurelien Henry, *Seven Ways of the African Warrior*, Republic of Cameroon, Shiai Magazine, April 2008

Pressfield, Steven, *The Warrior Ethos*, New York, N.Y., Black Irish Entertainment, LLC, 2011

Sun-tzu, *The Art of War*, translation by Ralph D. Sawyer, New York, N.Y., Barnes & Noble Books, Inc., 1994

Taylor, Colin F., *Native American Hunting and Fighting Skills*, Guilford, CT., The Lyons Press, 2003

Thucydides, *The Peloponnesian War*, translation by Richard Crawley, edited by Robert B. Strassler, published in *The Landmark Thucydides: A Comprehensive Guide to the Pelopponesian War*, New York, N.Y., The Free Press, 1998

U.S. Marine Corps, *Fleet Marine Force Manual 1, Warfighting*, Washington, D.C., U.S. Government Printing Office, 1989

Valadez, David M., *"Define 'Warrior,'"* blogpost, senshincenter.com/blog, Goleta, CA., Senshin Center, 2022

Zeilinger, Ron, *Lakota Life*, Chamberlain, S.D., Tipi Press Printing, 1986

Bibliographic Notes:

As far as *Beowulf* is concerned, there is no single, universally agreed-upon translation of the epic poem. Nor do the virtues from *Beowulf* listed in Chapter Four appear in the text as a list. Some of these virtues are stated, and others are extrapolated from the textual meaning of the story. Scholars may debate the issue, but I've chosen to enumerate in a list the Beowulfian warrior virtues in order to get the point across to the reader.

I obtained a copy of Shannon E. French's book *The Code of the Warrior* to assist me as a source during my research for this book. I had completed the first draft of this book before I ever looked at it. I was struck by how similar the content of her writing on this subject is to mine, which speaks to the universality of warriors and warrior codes. Dr. French's work is an academic treatment of warrior codes, a work of history and philosophy. The purposes of our two books are different, but I highly recommend Dr. French's work to you, should you desire an in-depth, academic treatment of warrior codes.

In *First to Fight, An Inside View of The U.S. Marine Corps*, General Krulak consistently uses the word "fighter" to describe Marines. If one were to cross out the word "fighter" and insert the word "warrior," it works nicely. Note that the title of Krulak's book is "First to Fight," not "First to Warrior," so the use of the word "fighter" is consistent. If you are a U.S. Marine, General Krulak's book is a "must read."

The quoted verses in this volume are from noted thinkers, leaders, jurists, warriors, writers, and philosophers. With the exception of General Mattis and Major Allred, all are now dead, leaving these kernels of wisdom as their legacies to us. While some of them do have their detractors, if you have never heard of them, I recommend learning more about them.

Lastly, this work is the product of the author's mind and is original work. It is based on the author's education, training, experience, and research. No system of Artificial Intelligence was utilized in the writing of this book.

RECOMMENDED READING

I recommend the following books to you:

Living the Martial Way, by Forrest E. Morgan. This is a fantastic book; however, I do not agree with everything he says. He is a very traditional martial artist (and a now-retired U.S. military officer) who has studied several Japanese martial arts systems. For example, he states that judo is not a martial art. I understand what he is saying but I believe that to be a preposterous assertion. Overall, though, his book is an excellent guide on how to live like a warrior, and I list it here first for a reason.

Another is *The Warrior Ethos* by Steven Pressfield. Mr. Pressfield is a great writer. He is also a former U.S. Marine. This volume is very much worth reading. Some critics have argued that Pressfield's presentation of the warrior ethos is anachronistic, so you be the judge. Even if it is, it is still very much worth reading.

Two other recommended books by Pressfield are *Gates of Fire* and *Tides of War*. These are works of historical military fiction and are very instructive as to the warrior archetype. These books tell great stories and are entertaining to read.

Do not discount fiction as part of your warrior reading. Fiction often delves into and examines the human condition in ways that non-fiction cannot. While works of non-fiction such as histories and biographies should be the primary basis for the warrior's reading list, fictional works should also be included for this reason.[38]

---

[38] Please see the Note on Reading Fiction on page 86.

INDEX

This index covers Chapters One through Nine. Other book sections are excluded.

A NOTE ON FAITH

As I mention in the book, the warrior's source of strength lies in the warrior code and the warrior's faith system, family, and friendships. This book is not designed to proselytize or evangelize or advance any religious agenda. But it is important to note that a warrior ought to believe in something greater than him or herself.

Just as warriors must *serve* something greater than themselves, they ought to *believe in* something greater than themselves. I use "ought" instead of "should" because faith is deeply personal, and I am in no position to preach to you. The fact that this entire book might be considered nothing more than me preaching about warrior attributes and warrior codes is an irony not lost on me here. But I am not preaching on religion; that's for sure.

I once knew a woman, an Olympic-level athlete and U.S. Army soldier, who was smart, attractive, great at her job, a wonderful person, and a total badass who didn't believe in any higher power or higher level of existence. If women were allowed to attend Ranger school in her era, she would have solidly passed it. I admired her a great deal.

She did not believe in any higher power or a life hereafter. You may call such a higher power or existence God, or The Universe, or Mother Nature, or Mother Earth, or whatever, but she believed in nothing like that. What she believed in was herself. "I believe in *myself*," is what she would say.

She also read self-help books and followed the teachings of certain self-help gurus. She was constantly seeking to improve herself (there's nothing wrong with self-improvement but follow me here). She was constantly seeking "the answer" from other people – *other humans*, just like *her*, humans who were *selling* "the answer." She was a seeker. She was a seeker because, in my opinion, she was not truly confident in her *belief* in herself.

You see, you are not big enough for you. Let me say that again, in exactly the same way: You... are... not... big... enough... for... you. You can believe in yourself, but as a human being, you know, intrinsically, that you have limitations. Your belief in yourself only goes as far as your innermost fear or innermost lack of confidence.

I've known big, strong, badass men who are terrified of public speaking. *Everyone* has an innermost fear(s), an innermost lack of confidence, in *something*. You are not big enough for yourself. You ought to have faith in something outside of yourself.

Whether that faith lies in religion or philosophy, whether you call that higher power "God" or "The Great Spirit," or that higher plane of existence "The Universal Consciousness" or something else, as a warrior it will benefit you to believe in something greater than yourself.

Such a belief can help provide emotional and, yes, spiritual sustenance in the dark times. As a human being, you will certainly experience dark times. As a warrior, you will experience dark times.

I choose the Christian faith, but you should follow your own path. Let your conscience be your guide. Do not leave this aspect off of your warrior path. Faith in something greater than yourself will provide light for your path.

You will need that light.

A NOTE ON LEARNING SELF-SUFFICIENCY

Start with building and maintaining physical strength and conditioning. Eat whole foods that do not use hormones, are not genetically modified, and where no excessive pesticide application has been used, whenever possible.

Obtain your foods from locally sourced farms or grow your own. Such health and fitness maintenance will assist you in developing self-sufficiency.

Also learn hard skills. Hard skills include mechanics, welding, carpentry, masonry, plumbing, electrical, gardening, and anything to repair a vehicle or a home, or to build or repair useful things.

Learn self-defense in some form, from boxing, wrestling, or the other martial arts, to weapons and firearms. This includes safe firearms handling, utilization, and storage. Learn the laws applicable to self-defense in your area.

Build a strong social network that is local and is not one-dimensional. Your "gun bros," "gym bros," and "mat bros," or any other group you call "bros," are not sufficient. A beneficial social network should have diverse people, interests, skills, and resources.

In the long run, skills and experiences are more valuable than material possessions or wealth. Knowing how to do things for yourself is wealth in itself.

A NOTE ON READING FOR U.S. MARINES

It is essential that, among your selection of reading materials, you read two books, most particularly if you are – or plan to be – an officer. Those books are *First to Fight, An Inside View of the United States Marine Corps* by LtGen Victor H. "Brute" Krulak, and *With the Old Breed at Peleliu and Okinawa* by E.B. Sledge, who fought with the 1st Marine Division in the Pacific Theater, World War II. These are *must-reads*. Semper Fi.

A NOTE ON THE NATURE OF THE WARRIOR AND WAR (FORCE AND VIOLENCE)

The warrior may be called upon to protect and defend others. The warrior may have to use force to protect and defend. The warrior may be called upon to *en*force society's rules on fellow members of society. Encapsulated in the word "enforce" is the word "force."

The warrior uses force judiciously and fairly, not arbitrarily or capriciously. While force may have to be applied with a level of violence in order for it to be effective, it is only applied when necessary and at the level required by the circumstances. It is never applied "just because" or "because he deserved it." It is applied in order to achieve a societal good. For example, the arrest of a resisting, violent criminal may require force, and it may be that in that circumstance such force must be applied violently in order for that force to be effective.

Force applied violently when it is necessary to be applied violently – when the use of such force is *authorized* by the law or the rules of engagement, and when it is *necessary* to apply it with violent action – is not dishonorable to the warrior.

Violence, however, for its own sake, such as is used by a criminal in order to advance his criminal self-interest, is dishonorable. Strong people who enjoy hurting weaker people, just for the fun of it, are dishonorable and are not warriors. Such persons use force and violence not for a greater purpose, or greater good, but because they enjoy hurting others and/or to further their own self-interest.

Force and violence are not the same thing, at least not in the context of warriorship. Force may require violence to achieve its ends for the greater good. But violence in and of itself, used to advance self-interest, is simply wrong. Force is a tool for good; violence is a tool for evil.

Sometimes the warrior is called upon to be a destroyer in the protection of others, is required to destroy human life and destroy human creations. As distasteful as this is, sometimes it is absolutely necessary to do so to protect one's society, and/or to protect humanity at large.

It would be great if there were no more Adolf Hitlers, no more Josef Stalins, no more Mao Zedongs, no more Pol Pots or Vladimir Putins. But there are such people in our world, and there always will be. Sometimes they must be stopped for the good of humanity. It is the warrior who is called upon to stop them.

There is no dishonor in being the instrument of destruction when it is absolutely necessary for the survival of a society or of humanity. This is the warrior's role.

A NOTE ON READING FICTION

As discussed in Chapter Eight, a warrior should read about warriorship. Reading furnishes various perspectives on warriorship and arms the warrior reader with greater depth of perspective on the world. Non-fiction should be the basis for warrior reading, but as I mentioned, fiction should not be ignored, as it often deals with the human condition in ways that non-fiction cannot.

But what kind of fiction? My recommendation is to read historical military fiction and military science fiction. If sci-fi doesn't power your rocket ship, that's okay. Stick to historical fiction.

For me, reading military science fiction is beneficial because stories can present moral, ethical, strategic, tactical, and logistical challenges in ways that are unreachable when writing a non-fiction history or biography. A fiction author is free to develop scenarios within stories that require innovative solutions by the characters in those stories, which can be instructive.

Military science fiction can also present thorny leadership challenges that would not necessarily be replicated in real life, but which can be instructive as to decision-making and the leading of people.

If you enjoy military science fiction, or want to get started in it, I recommend the *Lost Fleet* series by Jack Campbell. There are many others; *Ender's Game* by Orson Scott Card is another great one. You just have to look.

A SPECIAL NOTE ON THE INTERNET AND SOCIAL MEDIA

Social Media:

In this, the so-called 'digital age," we are all confronted by an avalanche of information found on and through the internet and social media. The internet and social media are great communication tools, and the internet has made research a relatively simple task as opposed to extensive brick-and-mortar library research of days past. But, as in all things, the benefits are but one side of the coin. On the other side of the coin is the dark side – the side that is harmful, not helpful.

Social media, in particular, is the province of "provisional reality." Provisional reality is something that a social media user specially curates to present to the world. Provisional reality provides a "reality" that is not necessarily real at all.

Sometimes, the user simply wants to portray themselves in a certain, very particular way for the audience. A sort of "Hey, look at me, I'm great" presentation. For others, it is in the pursuit of commercial goals, such as to become an "influencer" in order to make money. Often, these pursuits overlap, as do the presentation techniques.

Do not be fooled. When you see someone on social media who portrays themselves as having a perfect life, of achieving great things, whether commercial success, fame, or other tangible gain with little or minimal effort, do not believe it. No one, and I mean no one, has a perfect life. What you see on the outside is just what the person wants you to see; everything else, the true reality, is hidden from view.

The problem here is that many, many people compare themselves to persons they see on social media. If someone portrays a perfect life to you on social media and you despair that you do not have a perfect life, well, join the club. No one has a perfect life. Even if it looks like they do, they do not. Absolutely no one on this planet has a perfect life.

Be careful of what you see in celebrities, social media influencers, professional athletes, and the like on the internet, because you are being furnished a version of their lives that is created for public consumption. "Created," and its companion word, "curated," are very nice words that mean, "They made it up." Don't be fooled.

Social media, which can also be described fairly as "anti-social media" considering its negative influence and impact, is the province of people who subsist and live on the outrage of the day. These people are loaded, cocked, and ready to fire with criticism, condemnation, and complaints at whomever and whatever outrages them at any given moment in time.

These outraged denizens of the various social media platforms do nothing to benefit humanity and purvey moral and psychological poison to the population. The warrior is not fooled and does not fall prey to such poison.

The warrior, as a leader, strives to say only what is honest, true, and kind, and is measured in his or her interaction with others. The warrior seeks to be a benefit to humanity.

Nothing you see on the internet and social media should cause you to feel bad. Remember, almost all of the "wonderful" and "perfect" people you see on social media have simply made up how wonderful and perfect they are. Don't be fooled.

Artificial Intelligence:

Going forward into the twenty-first century, you will be confronted by extensive artificial intelligence (AI) and corresponding "deep fakes." A "deep fake" is a doctored piece of media, such as a video, of a celebrity or political leader doing or saying something that they actually neither did nor said.

Imagine a live White House press conference being held where the president makes an official address and says, "I have just ordered a thermonuclear attack on Country X. The missiles are on the way now."

How, and how quickly, will Country X respond in kind? This is sort of a problem, a really big sort of problem, if this address never actually happened but is the product of an AI-driven deep fake posted by a malevolent person, group, or country.

This is the future you will have to navigate. Artificial intelligence will make knowing truth from fiction very, very difficult. Work very hard to avoid being fooled. Always work to find and know the truth.

Treat everything that you see on the internet and social media with some skepticism. Make that skepticism habitual so that you won't be fooled.

In no way should you let anything or anyone you see on the internet and social media make you sad, unhappy, or depressed. Nothing you see there should lead you to feel bad or do anything bad. Much, if not most, of what you see there is simply made up. Don't be fooled.

Remember your code: your moral, ethical, and martial compass, as you walk your Warrior Path. It will help keep you from being fooled by providing you with a solid foundation of belief.

RESOURCES FOR VETERANS

Veterans Crisis Line

The Veterans Crisis Line connects veterans and service members in crisis and their families and friends with qualified, caring VA responders through a confidential toll-free hotline, online chat, or text.

- **Dial 988 and press 1**, 24/7.
- *For non-veterans, dial 988.*

War Vet Call Center

Call **1-877-927-8387**, 24/7 to reach the Vet Center Call Center and talk about your military experience or other concerns during your transition from military to civilian life. The team is comprised of veterans from several eras as well as family members of veterans.

Women Veterans Call Center

The Women Veterans Call Center (WVCC) provides VA services and resources to women veterans, their families, and caregivers. You can also chat online anonymously with a WVCC representative.

- Call **1-855-VA-Women** (**1-855-829-6636**) 8 a.m.–10 p.m. ET, Monday–Friday;8 a.m.–6:30 p.m. ET, Saturday.

Real Warriors

Real Warriors, a program through the Defense Centers of Excellence for Psychological Health and Traumatic Brain Injury (DCoE), provides information and resources about psychological health, Post-Traumatic Stress Disorder (PTSD), and traumatic brain injury.

- Call **1-866-966-1020**, available 24/7.

Caregiver Support

If you are caring for a Veteran, the VA Caregiver Support Program offers training, educational resources, and a variety of tools to help you succeed.

- Call **1-855–260–3274**, 8 a.m.–8 p.m. ET, Monday–Friday, for advice on being a caregiver.
- Find your local Caregiver Support Coordinator.

Coaching Into Care

This VA program provides guidance to veterans' family members and friends for encouraging a reluctant veteran they care about to reach out for support with a mental health challenge.

- Free, confidential assistance is available by calling **1-888-823-7458** Monday–Friday, 8 a.m. –8 p.m. ET, or emailing CoachingIntoCare@va.gov.

AFTERWORD

It is my hope that this work adds favorably to the discourse on warriors and warriorship. It is also my hope that this work helps someone, somewhere, in a positive way. It was written by a "regular Joe" who just happens to have had some experiences that inform his point of view on this subject, and who believes that he should pass it on to those who may need it.

ABOUT THE AUTHOR

Todd K. Hulsey served thirty-nine years of regular, reserve, and guard military service. He entered active duty with the United States Marine Corps as a private in late 1981, retiring from the United States Air Force Reserve in early 2010. Entering the Texas State Guard in 2014, he retired from the Guard in 2019 as the last wing commander of the 4th Air Wing, whose colors were permanently cased on December 31, 2019. The bulk of his military service was in Intelligence and Defense Support of Civil Authorities billets.

He also served as a federal law enforcement officer for over twenty years, starting as a special agent with the U.S. Treasury Department in 1988 and retiring as a supervisory special agent of the Federal Bureau of Investigation in 2014. His service in uniform and as a federal agent took him to Asia, Australia, Europe, Latin America, the Middle East, and to various domestic duty locations.

Some of Colonel Hulsey's notable assignments include as FBI representative to the Sub-Policy Coordinating Committee for

Counterproliferation, National Security Council, from 2004-2005 under NSC director David Stevens during the George W. Bush administration, and as a senior FBI representative on detail to the Central Intelligence Agency, Directorate of Operations, from 2005-2007.

In 2005, he was selected to help draft the National Implementation Plan for the War on Terror, which was approved by President George W. Bush in 2006. He and his fellow strategic planners were awarded a Meritorious Unit Citation by Vice Admiral (Ret.) Scott Redd, director of the National Counterterrorism Center, for their work developing the plan.

In 2009, he served as the Assistant Legal Attaché for International Terrorism at the U.S. Embassy in Riyadh, Kingdom of Saudi Arabia.

He is U.S. Intelligence Community Joint Duty Accredited, is a Certified FBI Intelligence Officer, and a Certified FBI Counterintelligence Officer. He is a graduate of the FBI

Executive Development Institute, the Air Command & Staff College, and the Defense Nuclear Weapons School.

When you are finished with this book, we hope that it finds a home on your bookshelf as part of your Warrior Library. If you do not intend to keep it, we recommend the following:

Give this book to someone who you think would want, or need, to read it.

Donate it to your local library.

Donate it to a charity.

Sell it to a used bookstore.

Place it in a recycling bin.

Use, reuse, repurpose, recycle.

Thank you for reading this book.

www.ingramcontent.com/pod-product-compliance
Lightning Source LLC
Chambersburg PA
CBHW071902020426
42331CB00010B/2639